MATH ON ✕ ➖ ➕ ➗ MY MIND

The LONG and SHORT OF IT

EL LARGO Y CORTO DE ÉL

Bilingual Edition English-Spanish | Edición bilingüe inglés-español

Karen Clopton-Dunson

EZ Readers lets children delve into nonfiction at beginning reading levels. Young readers are introduced to new concepts, facts, ideas, and vocabulary.

EZ Readers permite que los niños se adentren en la no ficción en los niveles iniciales de lectura. Los lectores jóvenes son introducidos a nuevos conceptos, hechos, ideas y vocabulario.

Tips for Reading Nonfiction with Beginning Readers
Consejos para leer no ficción con lectores principiantes

- Begin by explaining that nonfiction books give us information that is true.
- Comience explicando que los libros de no ficción nos dan información verdadera.
- Most nonfiction books include a Contents page, an index, a glossary, and color photographs. Share the purpose of these features with your reader.
- La mayoría de los libros de no ficción incluyen una página de Contenido, un índice, un glosario y fotografías en color. Comparta el propósito de estas características con su lector.

- The **Contents** displays a list of the big ideas within the book and where to find them.
- **El Contenido** muestra una lista de las grandes ideas dentro del libro y dónde encontrarlas.
- An **index** is an alphabetical list of topics and the page numbers where they are found.
- Un índice es una lista alfabética de temas y los números de página donde se encuentran.
- A **glossary** contains key words/phrases that are related to the topic.
- Un glosario contiene palabras clave o frases relacionadas con el tema.
- A lot of information can be found by "reading" the **charts** and **photos** found within nonfiction text.
- Se puede encontrar mucha información al "leer" los cuadros y las fotos que se encuentran en el texto de no ficción.

With a little help and guidance about reading nonfiction, you can feel good about introducing a young reader to the world of *EZ Readers* nonfiction books.

Con un poco de ayuda y orientación sobre la lectura de no ficción, puede sentirse bien al presentar a un joven lector al mundo de los libros de no ficción de *EZ Readers*.

First Edition, 2020.

Author/Autor: Karen Clopton-Dunson
Designer/Diseñador: Ed Morgan

Names/credits:
Title: The Long and Short of it El Largo y Corto de Él / by Karen Clopton-Dunson
Description: Hallandale, FL : Mitchell Lane Publishers, [2020]

Series: Math on My Mind

Library bound ISBN: 9781680205473
eBook ISBN: 9781680205480

EZ Readers is an imprint of Mitchell Lane Publishers

Bilingual Edition English-Spanish
Edición bilingüe inglés-español

Photo credits: Freepik.com, Shutterstock.com

CONTENTS
CONTENIDO

Matthew has lots of questions:
How long is his toy car?
How wide is his book?
How tall is his robot?

Mateo tiene muchas preguntas:
¿Qué largo es su carro de juguete?
¿Qué tan ancho es su libro?
¿Qué altura tiene su robot?

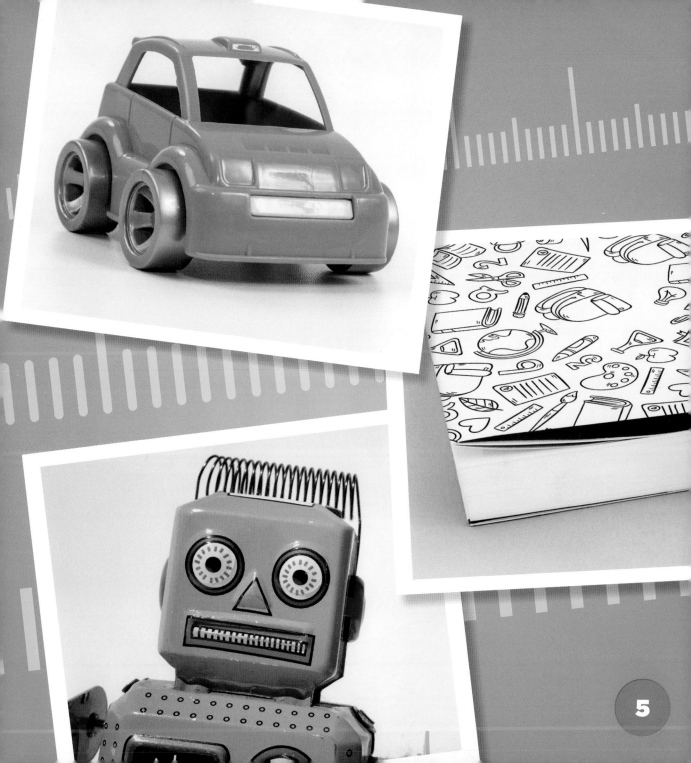

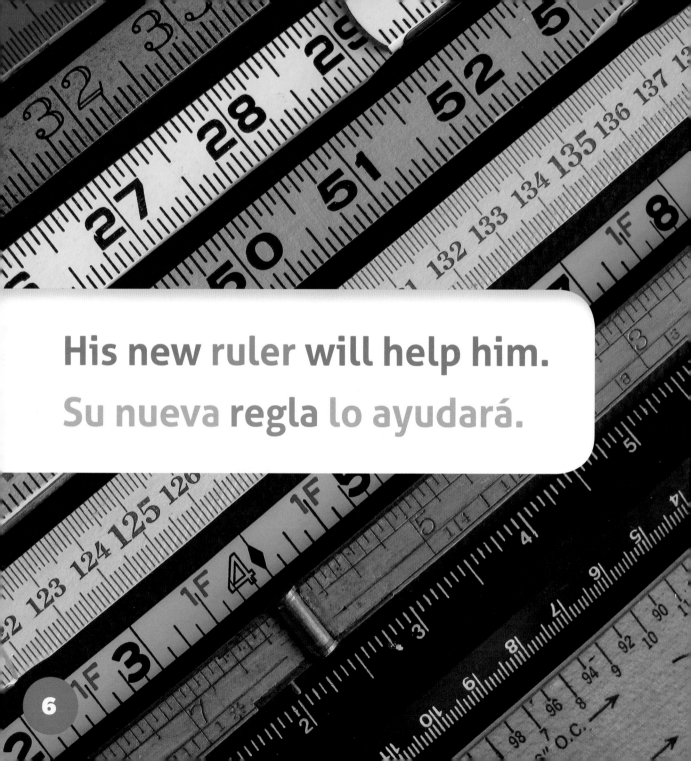

His new ruler will help him.

Su nueva regla lo ayudará.

6

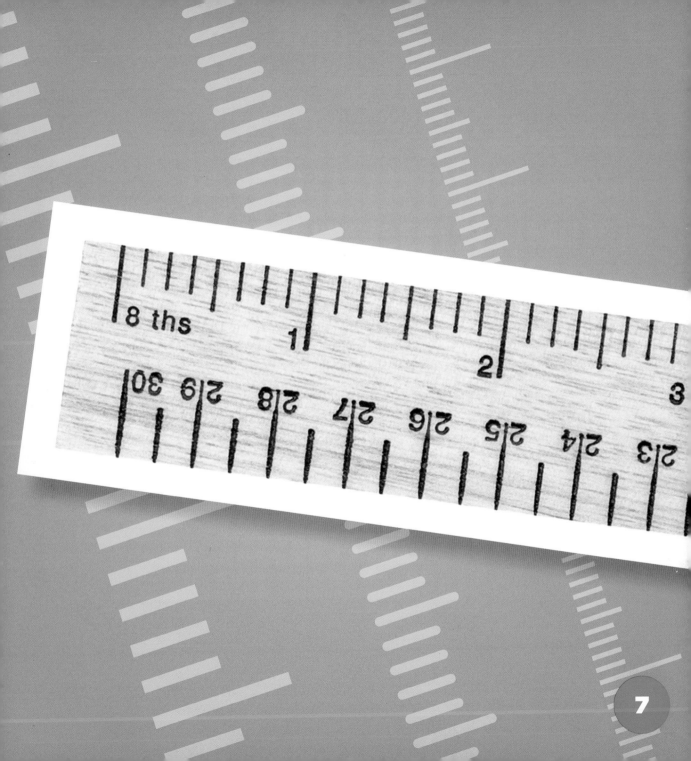

A ruler is a measuring tool.
It shows the numbers 0 to 12.
Sometimes the zero is not seen.

Una regla es una herramienta
 para **medir**.
Muestra los números del 0 al 12.
A veces no se ve el cero.

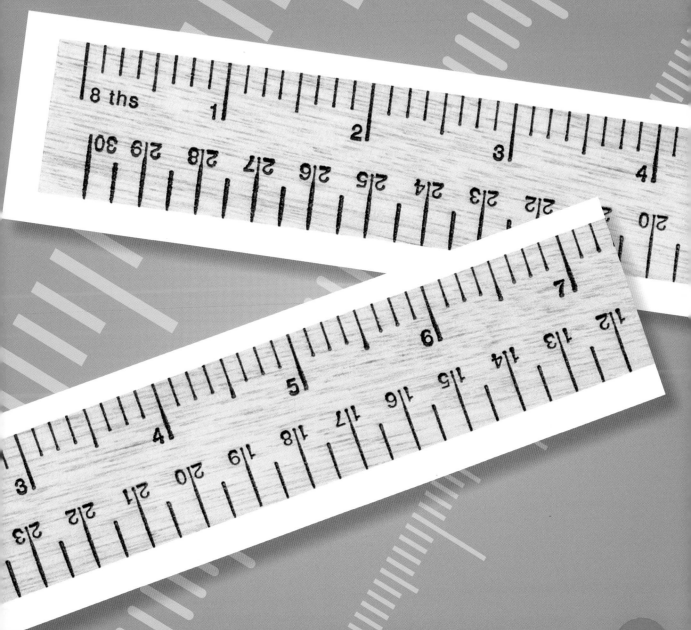

Matthew lines up the zero inch mark with the left side of the toy car.
The car ends at the number three. It is three inches long.

Mateo alinea la marca de cero pulgadas con el lado izquierdo del carro de juguete.
El carro termina en el número tres. Tiene tres **pulgadas** de largo.

Matthew's book is on the table.
He sets the ruler across the book.
He makes sure the zero edge of the ruler
 and the left side of the book line up.
His book is five inches wide.

El libro de Mateo está sobre
 la mesa.
Él pone la regla a través del libro.
Se asegura de que el borde cero de
 la regla y el lado izquierdo del libro
 estén alineados.
Su libro es de cinco pulgadas de ancho.

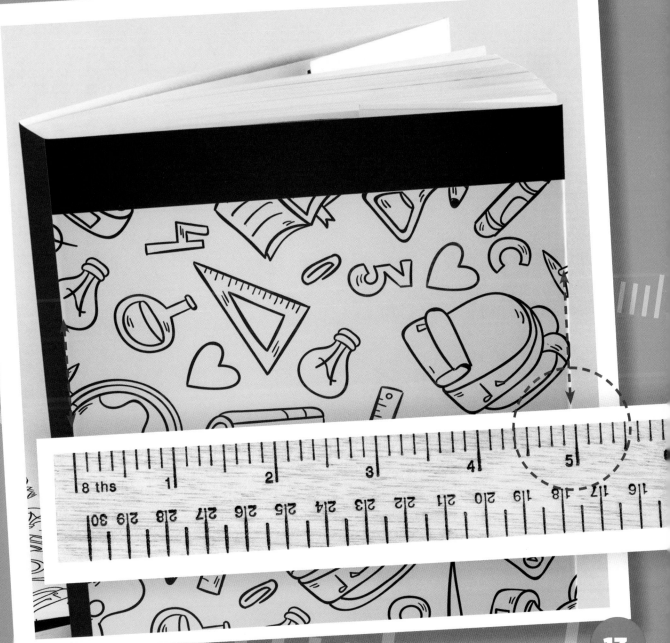

He wants to see how tall his robot is.
He measures from the foot to
  the head.
The zero end of the ruler is by
  the robot's foot.

Quiere ver qué tan alto es su robot.
Él mide desde el pie hasta la cabeza.
El cero, al final de la regla esta por
  el pie del robot.

The robot is seven inches tall.

El robot mide siete pulgadas de alto.

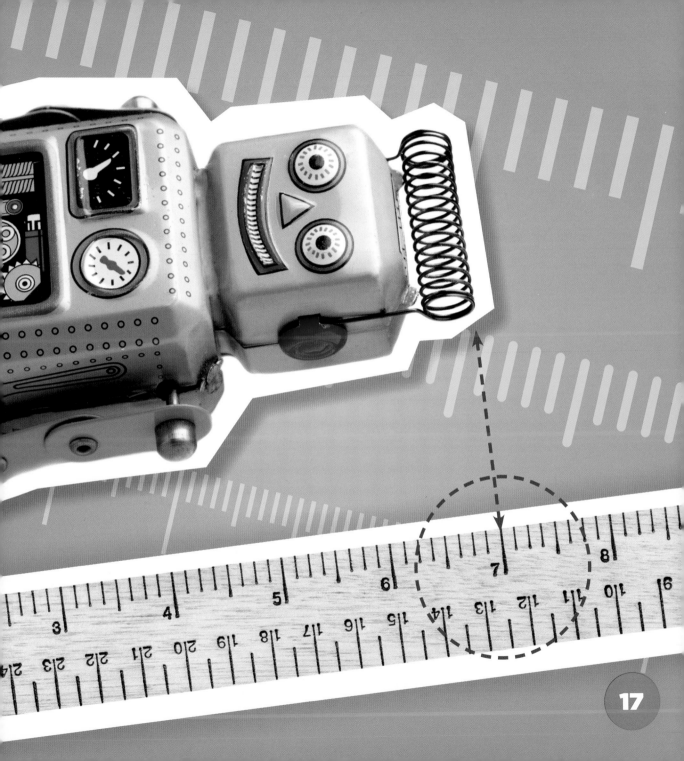

He turns the robot on.
Away it goes!

Enciende el robot.
¡Se va!

Measuring time is over.

El tiempo de medir ha terminado.

Measuring Activities Actividades de Medir

Line up several stuffed animals of varying heights. Arrange them from the tallest to the shortest.
Alinear varios animales de peluche de diferentes alturas. Organízalos desde el más alto al más corto.

Have children connect Unifix Cubes. Now guess how many cubes it would take to measure the length of a wooden spoon, pencil, book, etc. Write down the estimate, then use the cubes to measure each object. Write it down. Compare the estimate with the actual measurement. Ask the students if they guessed correctly.
Haga que los niños conecten los cubos Unifix. Ahora adivina cuántos cubos se necesitarían para medir la longitud de una cuchara de madera, un lápiz, un libro, etc. Anota el estimado y luego usa los cubos para medir cada objeto. Escríbelo. Compara el estimado con la medida real. Pregunte a los alumnos si adivinaron correctamente.

Ask children to use their ruler to measure the length of toys. Write down the inches for each one. Which toy is the longest?
Pídales a los niños que usen su regla para medir la longitud de los juguetes. Anote las pulgadas para cada uno. ¿Qué juguete es el más largo?

Glossary Glosario

edge The line or part where an object or area begins or ends
borde La línea o parte donde un objeto o área comienza o termina

inches A unit of measurement equal to 1/36 yard or 1/12 of a foot (2.54 centimeters)
pulgadas Una unidad de medida igual a 1 / 36 yarda o 1 / 12 de un pie (2.54 centímetros)

long Extending a great distance from one end to the other end: not short
larga Distancia de una gran distancia de un extremo a otro: no corta

measure Something (such as a cup or a ruler) that is used to measure things
medir Algo (como una taza o una regla) que se usa para medir cosas

ruler A straight piece of plastic, wood, or metal that has marks on it to show units of length and that is used to measure things
regla Una pieza recta de plástico, madera o metal que tiene marcas para mostrar unidades de longitud y que se usa para medir cosas

tall Having a specified height
alto Teniendo una altura especificada

wide Extending a great distance from one side to the other: not narrow
ancho Extendiéndose una gran distancia de un lado a otro: no estrecho

Further Reading Otras Lecturas

Cleary, Brian P. *How Long or How Wide.* January 2014.

Heos, Bridget. *Measure It.* September 2014.

Higgins, Nadia. *Math It! Measure It!* August 2016.

Weakland, Mark Andrew. *How Tall? Wacky Ways to Compare Height.* July 2013.

On the Internet En Internet

Free Online Measurement Games | Education.com
https://www.education.com/games/measurement/
Uses interactive Measurement Games to reinforce understanding of lengths and heights

33 Measurement Games - Educational Fun Activities for Kids Online
https://www.splashmath.com/measurement-games
Looking for educational fun activities for students / kids to help them learn math? Splash Math offers cool interactive problem-solving Measurement Games.

Measurement Games | Turtle Diary
https://www.turtlediary.com/games/units-of-measurement.html
Measurement games designed for kids to teach them how to measure in a fun-filled way.

Measure It! - a game on Funbrain
https://www.funbrain.com/games/measure-it
Measure length in centimeters or inches. Choose the measurement that matches the length of the red bar.

About the Author Sobre el Autor

Karen Clopton-Dunson has taught kindergarten and Head Start in the Chicago Public Schools. Helping children have fun learning math concepts is her reason for writing this book. She currently lives in the Chicagoland area. This is her second book with Mitchell Lane Publishers.

Karen Clopton-Dunson ha enseñado el jardín de infancia y un programa que se llama "Head Start" en las Escuelas Públicas de Chicago. Ayudar a los niños a divertirse aprendiendo conceptos matemáticos es su razón para escribir este libro. Actualmente vive en el área de Chicago. Este es su segundo libro con Mitchell Lane Publishers.